21st Century Junior Library

UNDERSTANDING FLOODS

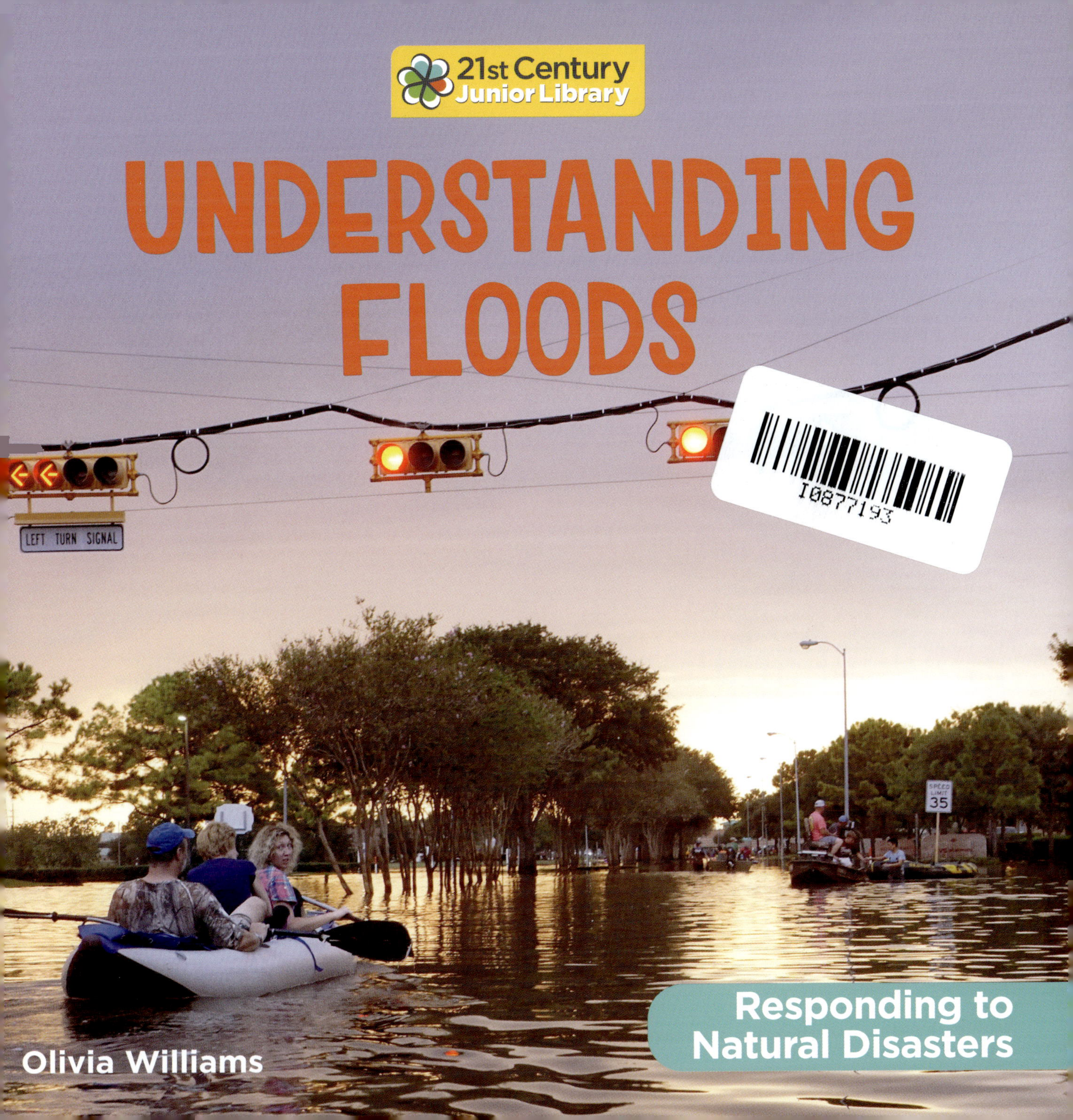

Responding to Natural Disasters

Olivia Williams

Published in the United States of America by:

CHERRY LAKE PRESS
2395 South Huron Parkway, Suite 200, Ann Arbor, Michigan 48104
www.cherrylakepress.com

Reading Adviser: Beth Walker Gambro, MS, Ed., Reading Consultant, Yorkville, IL
Photo Credits: © IrinaK/Shutterstock.com, cover, 1; © Arnold O. A. Pinto/Shutterstock.com, 5; © Sophon Nawit/Shutterstock.com, 6; © CPA Media Pte Ltd/Alamy Stock Photo, 7; © Maximillian cabinet/Shutterstock.com, 8; © northallertonman/Shutterstock.com, 9; © narongpon chaibot/Shutterstock.com, 10, 11; © Arthur Villator/Shutterstock.com, 12 [left]; © FashionStock.com/Shutterstock.com, 12 [right]; © Mikhail Malyshev/Shutterstock.com, 13 [top]; © IrinaK/Shutterstock.com, 13 [bottom, left]; © sunakri/Shutterstock.com, 13 [bottom, right]; © jps/Shutterstock.com, 15 [top]; © Max Lindenthaler/Shutterstock.com, 15 [bottom]; © weerastudio/Shutterstock.com, 16; © Hank Erdmann/Shutterstock.com, 19; © Simone Hogan/Shutterstock.com, 20

Cherry Lake Press is an imprint of Cherry Lake Publishing Group.

Library of Congress Cataloging-in-Publication Data
Names: Williams, Olivia (Writer of children's books), author.
Title: Understanding floods / Olivia Williams.
Description: Ann Arbor, MI : Cherry Lake Publishing, [2022] | Series: Responding to natural disasters | Audience: Grades 2-3
Identifiers: LCCN 2021043062 | ISBN 9781534199187 (hardcover) | ISBN 9781668900321 (paperback) | ISBN 9781668901762 (pdf) | ISBN 9781668906088 (ebook)
Subjects: LCSH: Floods—Juvenile literature.
Classification: LCC GB1399 .W48 2022 | DDC 363.34/93—dc23
LC record available at https://lccn.loc.gov/2021043062

Cherry Lake Press would like to acknowledge the work of the Partnership for 21st Century Learning, a Network of Battelle for Kids. Please visit http://www.battelleforkids.org/networks/p21 for more information.

Printed in the United States of America
Corporate Graphics

ABOUT THE AUTHOR

Dr. Olivia Williams is from mid-Wales in the United Kingdom. She has traveled to many countries to help people after natural disasters. She has felt earthquakes and aftershocks, and seen damage from floods, typhoons, and wildfires. The natural environment can be harsh but also beautiful. Olivia loves to hike and run in the mountains and photograph the incredible things she sees.

CHERRY LAKE PRESS

CONTENTS

WHAT ARE FLOODS?

A flood occurs when too much water builds up in one area. The five main types of floods are coastal, river, flash, groundwater, and sewer. Occasionally, different types of floods happen at the same time in the same area.

Coastal floods can happen during a high tide. Storms like hurricanes can cause coastal floods too. River floods occur when an area gets a lot of rainfall.

Floods can happen anywhere in the world.

This causes the river to flow over the riverbank. Some cities are close to large rivers. This means they are **vulnerable** to flooding. Flash floods happen very quickly after heavy rain. The ground can't **absorb** the water fast enough.

A heavy truck or SUV can be carried away by as little as 2 feet (0.6 meters) of rushing water! And smaller, compact cars can be swept away in as little as 1 foot (0.3 m).

One of the worst floods in history happened in 1931 in China. About 140,000 people drowned immediately after the flood. An estimated total of 4 million lives were lost from the effects of the flood.

Groundwater floods occur when rain adds to the water already in the ground. That water rises back up through the ground and causes flooding. Unlike flash floods, groundwater floods can happen days, weeks, or even months after heavy rain. Sewer floods happen when water from the drains and sewers overflows. This can be very smelly and unhealthy! Sewers are full of dangerous bacteria that can harm us.

Other events in the environment can cause floods. For example, melting **snowpacks** and **glaciers** lead to flooding. Melting ice and bad weather events happen more often now than in the past. This is due to climate change.

According to scientists, snowpacks are melting earlier and earlier. A snowpack is the total amount of snow and ice that's on the ground.

Think!

Experiencing a flood can be very upsetting. Imagine your friend's home has been flooded. You notice that they're upset. What would you say to make them feel better? Can you think of any positive outcomes from the flood?

Floods are one of the most expensive disasters, as they cost a lot of money to repair damages.

HOW FLOODS AFFECT THE PLANET AND PEOPLE

Floods are the deadliest type of weather event in the United States. They have shaped U.S. history. Some Americans have had to move to states with a lower risk of floods. In 2005, Hurricane Katrina hit the country's southeastern coast. There was so much rainwater that the Mississippi River flooded. People died, homes were destroyed, and essential materials needed to survive were lost. The water

also caused power outages and polluted drinking water. These things forced people to escape to safety.

Across the world, around 21 million people are affected by floods each year. Some of the worst affected countries are India, Pakistan, Egypt, Iraq, China, and Brazil. The Yellow River in China is famous for its floods. Floods there have killed millions of people.

Floods can spread disease. Playing in a flooded area might seem fun, but it's not safe! Stay above water or wear boots.

Floods do not only harm people. They also hurt animals and vegetation. They harm natural **ecosystems**. The habitats of small wildlife animals can be destroyed. In big floods, even large animals can be swept away and killed. **Pesticides** used in farming crops can also be washed into freshwater. Pesticides are unhealthy for humans, animals, and plants.

Make a Guess!

Low-income families are negatively affected by floods more than any other demographic. Can you guess why this is? Discuss your thoughts with a friend or family member. Ask them what they think.

Sandbags act as a wall and can help redirect the water.

PROTECTING AGAINST AND RECOVERING FROM FLOODS

Like other **natural disasters**, floods are difficult to prevent or fight. But there are basic things people can do to help. They can place sandbags against doors and put valuables on high shelves. When floods are severe, people are told to **evacuate**. When this happens, it is important that people take only necessary items. They should place medication and important documents like passports and birth certificates in plastic bags. People should pack enough food and water for their family and pets.

Malawi is a country in Africa. It often suffers from floods. This is because the rain from the mountains runs down onto the flat land. There are very few trees on the flat land due to **deforestation**. But trees can help to slow floodwater. Communities in Malawi are now planting trees to help protect their villages. Planting trees where there are none is called **afforestation**.

Technology can help too. Flood-related apps for smartphones can warn of local floods.

Create!

On a piece of paper, write out directions to a daily task. An example could be a household chore or a routine of getting ready for school. How might these tasks be affected by a flood?

Trees are like sponges. They can soak up excess stormwater.

Many companies and nonprofit organizations, including the American Red Cross, use technology to help people be better prepared for floods.

Automatic water barriers can stop water from entering a building. Solar lights help people see at night when floods have cut electricity. Technology can help communities before, during, and after a natural disaster.

Look!

Take a look around your home. Make a list of the things that are important to you. Are these things you can live without? Floods can damage homes and the things inside it. Now, take a look at the list you made. Cross off things you think you can live without. Discuss your list with a parent. Ask them what they'd put down on their list.

ACTIVITY

Which One Wins?

Did you know different types of soil are better at slowing down a flood? Gather a few different samples of soil and see which soil floods more quickly.

Materials Needed:

- 3 to 4 coffee filters
- 3 to 4 clear cups
- 3 to 4 rubber bands
- 3 to 4 different types of soil and sand
- Water

Instructions:

1. Place the coffee filters inside the top part of the clear cups.
2. Use the rubber bands to secure the edge of the coffee filters to the cups.
3. Add a good amount of soil or sand to each coffee filter. Make sure to use different soil or sand in each one.
4. Add water.
5. Watch and see which cup fills with water first!

GLOSSARY

absorb (uhb-SORB) to soak up liquid

afforestation (ah-for-uh-STAY-shuhn) planting trees on treeless land to create a forest

deforestation (dee-for-uh-STAY-shuhn) the action or process of clearing forests

demographic (deh-muh-GRAH-fik) refers to a particular section of a population or a specific group of people

ecosystems (EE-koh-sih-stuhmz) communities or groups of living organisms that live in and interact with each other in a specific environment

evacuate (ih-VAH-kyoo-wayt) to remove people from a place to take them somewhere safer

glaciers (GLAY-shuhrz) slowly moving masses of ice formed by compacted snow

natural disasters (NAH-chuh-ruhl dih-ZAH-stuhrs) natural events such as floods, earthquakes, or hurricanes that cause great damage

pesticides (PES-tuh-sydz) chemicals used to destroy plant or animal pests

snowpacks (SNOH-paks) masses of snow that have become hard under pressure

vulnerable (VUHL-nuh-ruh-buhl) capable of being easily hurt or injured, like small children, the elderly, or animals

FIND OUT MORE

Books

Abdo, Kenny. *How to Survive a Flood.* Minneapolis, MN: Abdo, 2019.

Raum, Elizabeth. *Flood!* Mankato, MN: Amicus High Interest, 2017.

Websites

Be Ready! Floods

https://www.cdc.gov/cpr/documents/beready_floods.pdf
Read the CDC's infographic on how to prepare in case of a flood.

Floods 101—National Geographic

https://www.youtube.com/watch?v=4PXj7bOD7IY
Learn more about floods.

INDEX